Space Weather

by Patrick H. Stakem

(c) 2018

Number 24 in the Space Series

Table of Contents

Introduction..3
Author..3
A note on Units..4
What is Space Weather?..4
Monitoring the Sun...6
    Ring Current...8
    Bow Shock..8
  How does Space Weather affect Earth?.....................9
    Coronal Mass Ejection...9
    Trapped Charged Particles....................................10
  How does it affect spacecraft?......................................11
    Spacecraft Charging..11
    Charged Particle effects...12
    Radiation Damage to electronics.........................14
    Mitigation Techniques..21
How does Space Weather affect the other planets?......22
    Venus..23
Mars..23
    Jupiter..23
Saturn...24
Uranus..24
Neptune...24
Pluto...24
How do we observe space weather?..............................25
    SDO...25
    Ulysses..25
    Parker Solar Probe..26
    RHESSI...28
Wrap-up..28
Glossary of Terms...30
Bibliography..34
Resources...37
If you enjoyed this book, you might also enjoy one of my other books in the Space series. 38

Introduction

This book covers the topic of weather in space, which is defined by our Sun. It interacts with our atmosphere and magnetic field (actually the atmosphere and magnetic fields (if any) of all the planets and moons), to influence our weather. It can impact our technology as well.

The study of Space Weather depends on space physics. All of our Space Weather originates with the Sun. The first serious study's of the subject were undertaken in the International Geophysical Year, in 1957. Actually stretching 18 months, this period saw the first launches of Earth Satellites, and amazing discovery's. We have to be able to predict Space Weather Storms, if we want humans to live and work in space. Space weather affects our satellites, produces the Northern (and Southern) Lights, and can affect communications, the electric grid, aviation and more.

Author

The author has a BSEE in Electrical Engineering from Carnegie-Mellon University, and Masters Degrees in Applied Physics and Computer Science from the Johns Hopkins University. During a career as a NASA support contractor from 1971 to 2013, he worked at all of the NASA Centers. He served as a mentor for the NASA/GSFC Summer Robotics Engineering Boot Camp at GSFC for 2 years. He taught Embedded Systems for the Johns Hopkins University, Engineering for Professionals Program, Graduate Computer Science for

Loyola University in Maryland, and Capitol Institute of Technology. He has done several summer Cubesat Programs at the undergraduate and graduate level.

A note on Units

I am fairly conversant in both English and Metric units (what is the metric equivalent of furlongs per fortnight?). Metric (SI) is mandated for NASA usage now, for interchangeability with our partner space faring nations. When a lot of the legacy flights discussed here were flown, English units were the norm. I have tried to keep the units comparable to the mission at the time. Conversions are easy enough, but units conversion is a source of error. It's not what you know about units and measurement, its how you think. And, I still think in English units (even the English use Metric now), and convert in my head or on my phone.

For scientific/engineering work, the Metric system is well thought out. For artisans, the English system served well, as most units were divisible by 2. Which is easy. Fold the cloth. Hopefully, when we are all taught Metric first, some one will still remember the conversions. You just need a good slide rule....

What is Space Weather?

The Sun controls the weather in our solar system. Large solar flares can release up to 10^{25} joules of Energy. The Sun releases electrons, stripped from their atoms, the resulting ions, and intact atoms from the corona. It also releases radio waves, which, traveling at the speed of light, reach Earth 8 minutes later. The particles travel at sub-light speeds. Bright aurora's in the polar regions will be created. Stellar flares have also been observed on other stars.

When the solar wind reaches the Earth's magnetic field, it interacts with it, creating a Geomagnetic storm. There can also be proton storms from the Sun. One effect is that the upper atmosphere is heated to tens of millions of degrees Kelvin. Luckily, it is near-vacuum. Increased electromagnetic radiation from radio to Gamma Rays are also observed.

NASA's *Wind* spacecraft was launched in 1994 to study radio waves and plasma in the solar wind. It continues to operate at this writing, from the Earth-Sun L1 point. WIND is a NASA mission to study radio waves and plasma. It was headed to the L1 Lagrange point, but along the way was used to study the magnetosphere and near-lunar environment in colloboration with the SOHO and ACE spacecraft. The spacecraft is currently operation as of this writing. It has 50 years of station-keeping fuel, but the electronics will give out long before then. It is operated from Goddard Space Flight Center.

A sister mission to WIND involves the Polar satellite, studying the polar magnetosphere and aurorae. It was launched in 1996, and returned data until 2008. It conducted multi-wavelength imaging of the aurorae, measured the entry of plasma into the polar magnetosphere and the geomagnetic tail, looked at the flow of plasma to and from the ionosphere, and determined the deposition of particle energy in the ionosphere and upper atmosphere. It had eleven science instruments. These included various electromagnetic field instruments, three sensors for particles, and three imagers, for the visible, ultraviolet, and x-rays.

As a side note, the out-streaming solar wind can be used with solar sails to catch a ride, in the same way a sailboat is moved by the wind on Earth. It is even possible to tack in towards the Sun. Solar sails are lightweight, and need to be reflective. It's not a fast way to go, but it will get you there eventually, without using fuel.

The Sun has an 11 year cycle, ranging between being fairly quiet, to very stormy. Solar flares are not completely understood, and there is no good model for their prediction. Generally, we have a 2-hour window between detection, to when the storm hits.

Monitoring the Sun

A lot of space missions were tasked with monitoring the Sun, to gather data that would lead to better models of its behavior. Between 1959 and 1968, four Pioneer Missions orbited at 1 A.U. to measure the Solar Wind and Magnetic Field. A 1970 mission, Helios, involved two spacecraft and the Apollo Telescope Mount on the then-orbiting Skylab. The Helios spacecraft went inside the orbit of Mercury. In 1980, the Solar Maximum Mission was launched to observe gamma rays, x-rays, and UV radiation. The author had flight software on that spacecraft. SMM failed in orbit (not my fault), but was retrieved and fixed by STS mission STS-41C. SMM reentered and burned in the atmosphere in 1989. My wife says, my best software burned with it.

The Japanese Yohkoh mission in 1991 observed solar activity such as X-ray flares. SOHO, the Solar and Heliospheric Observatory was a highly successful joint NASA-ESA mission.

The Solar Dynamics Explorer was launched in 2010, to the Lagrange point between the Earth and the Sun.

The Stereo mission in 2996 involved two identical spacecraft positioned to continuously fall behind and pull ahead of the Earth. They observe the same target (the Sun) from different positions, providing stereoscopic images.

As of this writing, the Indian Space Research Organization has a solar mission planned for 2021.

The Parker Solar Probe built at JHU-APL for GSFC, has just launched in a mission to monitor the Sun from a near position of about 8 solar radii. Why won't it burn up? It has a highly eccentric orbit, which allows it to cool off periodically. It also has a massive heat shield. (The old joke was, if you launch a solar mission, do it at night.) More details on Parker will be discussed later in the book.

The Ulysses probe was launched in 1990 to observe the unexplored solar polar regions. To get out of the plane of the ecliptic, Ulysses went to Jupiter for a gravity assist. At higher latitudes, the solar wind was slower than in the ecliptic plane, and there were large magnetic waves emitted. These scattered galactic cosmic rays.

NASA's Genesis was a sample and return mission to grab some of the solar wind for analysis. Most of the mission went well, but it crashed while trying to deploy a parachute to be snagged by a helicopter. It crashed into the Utah desert. Some of the samples were usable, however.

Ring Current

For a planet such as Earth, with a magnetic field, the Ring Current is caused by the movement of charged particles in the magnetosphere. It shields the lower latitudes from the magnetospheric electric fields. It extends about 3-8 Earth radii. The moving particles create a magnetic field, opposite to the Earth's. The ring current lies in the equatorial plane, and circulates clockwise. It shields the lower latitudes from electric fields in the magnetosphere. The ring current expands during geomagnetic storms.

Bow Shock

The bow shock is caused by plasma from the Sun hitting the Earth's magnetosphere. The same effects are observed at other planets with magnetic fields, Jupiter, particularly. The plasma is ionized, and follows spiral paths along magnetic field lines. The flow speed, at Earth, is around 400 km/s. The shock, at Earth, is some 17 km thick, and located 90,000 km Sun-ward. Bow shocks have been observed on planets in other star systems.

How does Space Weather affect Earth?

When a solar flare hits Earth's ionosphere, radio communications can be disrupted. The intensity of the Auroral displays over the poles is increased. Space weather in the vicinity of the Earth is dominated by the flare. Ground-based electrical transmission systems are affected, as the flares induce large currents in the wires.

Coronal Mass Ejection

A large solar flare occurred in September of 1859, and was observed by British astronomer R. C. Carrington in his private observatory on his estate outside of London. Both the associated sunspots and the flare were visible to the naked eye. The resulting geomagnetic storm was recorded by a magnetograph in Britain as well. They also recorded a perturbation in the Earth's ionosphere, that we now know was caused by ionizing x-rays. In 1859, this was all observed, but not understood. The ionosphere was not known to exist at the time. Now, we know a Coronal Mass Ejection from the sun, associated with a solar storm, is first seen as an energy burst hitting the Earth, and later by vast streams of charged particles, that travel slower than the speed of light. At normal levels, these particles cause the Northern and Southern lights.

What did happen, and was not immediately associated with the solar storm, was interference with the early telegraph systems of the time. The telegraph was relatively new, and wires stretched for many miles. Think of them as long antennas. The telegraph equipment was damaged, and large arc's of electricity started fires and shocked operators. No fatalities were reported. The employees of American Telegraph Company in New York found they could transmit messages with the batteries of their systems disconnected. The Northern lights were visible from Cuba. This was the largest such solar flare in at least 500 years ... so far.

What if such a super flare occurred today? First, we would have warning from sentinel satellites such as the Solar Dynamics Observatory, that are closer to the sun, and detect the passage of particles. They can tell us about this via radio, which travels faster than the particles. So, we would have a day or so's notice. All of our modern high-technology infrastructure would be at risk of damage, from the electrical grid to the Internet. Most of our satellites would be damaged, removing services we rely on such as long distance data communication, and navigation. It would be much better to turn everything off, and ride out the storm. Even that might not prevent major damage to networks. When is the next large solar event? Even the Astrophysicists can't tell us that. Only that it will eventually occur. Stay tuned...

Trapped Charged Particles

The van Allen belts are zones of charged particles, most of which come from capture of some of the solar wind. They are held in place by the planet's magnetic field. Mars, since it has no magnetic field, has no equivalent to the Van Allen Belt (and, incidentally, don't bother taking your compass). Below the Van Allen belts, there are fewer energetic charged particles. In the South Atlantic Ocean, the belts dip closer to the surface, so satellites might pass in and out of a charged particle environment every orbital period. This is called the South Atlantic Anomaly.

How does it affect spacecraft?

Space Weather impacts objects flying in space, including satellites. Some of these effects are discussed below. A Solar storm, with an increase in density and flow of the Sun's output will deform the Earth's magnetic fields, exposing more satellites to potential radiation damage. It also boils up the atmosphere, making it more dense at satellite altitudes. This causes more drag, and the satellites need to be re-boosted to a higher orbit.

In the case of very large solar flares, our sun-watching satellites will give us a few days warning. The astronauts in the International Space Station have a shielded "safe room" to ride out the storm.

Spacecraft Charging

An issue with on-orbit spacecraft is that they are not "grounded." This can be a problem when an electrical potential develops across the structure. Ideally, steps were taken to keep every surface linked, electrically. But, the changing phenomena has been the cause of spacecraft system failures. Where does the charge come from? Mostly, the Sun, in the forms of charged particles. This can cause surface charging, and even internal charging. Above about 90 kilometers in altitude, the spacecraft is in a plasma environment At low Earth orbit, there is a low energy but high density of the plasma. The plasma rotates with the Earth's magnetic field. The density is greater at the equator, and less at the magnetic poles. Generally, electrons with energies from 1-100 keV cause surface charging, and those over 100 keV can penetrate and cause internal charging. As modern electronics is very susceptible to electron damage, proper management of charging is needed at the design level.

Just flying along in orbit causes an electric field around the spacecraft, as any conductor traveling through a magnetic field does. If everything is at the same potential, we're good, but if there's a difference in potential, there can be electrostatic discharge. These discharges lead to electronics damage and failure, and can also cause physical damage to surfaces, due to arcing. This has been a problem at the International Space Station.

Charged Particle effects

On Earth, we are shielded from most radiation by our atmosphere and magnetic field. The Sun is the major source of our radiation, and high-energy particles. Right now, based on International Space Station experience, an Astronaut can have a maximum duration in space of about a year, before receiving his/her maximum lifetime dose. At the ISS you get, in a day, what some one on the ground would accumulate in a year. Shielding is one answer, but it can be counter-productive. Sometimes, a hit by a massively energetic particle can cause a spray of multiple lower energy particles, from the shielding itself.

There are two radiation problem areas: cumulative dose, and single event. Operating above the Van Allen belts of particles trapped in Earth's magnetic flux lines, spacecraft are exposed to the full fury of the Universe. Earth's magnetic poles do not align with the rotational poles, so the Van Allen belts dip to around 200 kilometers in the South Atlantic Anomaly. The magnetic field lines are good at deflecting charged particles, but mostly useless against electromagnetic radiation and uncharged particles such as neutrons. One trip across the Van Allen belts can ruin a spacecraft's electronics. Some spacecraft turn off sensitive electronics for several minutes every ninety minutes – every orbital pass through the low dipping trapped radiation belts in the South Atlantic. Spacecraft have been damaged or made inoperable by passage through the SAA. Recently, an International Space Station resupply ship was affected.

The Earth and other planets are constantly immersed in the solar wind, a flow of hot plasma emitted by the Sun in all directions, a result of the two-million-degree heat of the Sun's outermost layer, the Corona. The solar wind usually reaches Earth with a velocity around 400 km/s, with a density around 5 ions/cm^3. During magnetic storms on the Sun, flows can be several times faster, and stronger. The Sun has an eleven year cycle. A solar flare is a large explosion in the Sun's atmosphere that can release as much as 6×10^{25} joules of energy in one event, equal to about one sixth of the Sun's total energy output every second. Solar flares are frequently coincident with sun spots. Solar flares, being releases of large amounts of energy, can trigger Coronal Mass Ejections, and accelerate lighter particles like protons to near the speed of light.

Planets with magnetic fields will trap energetic particles arriving from the Sun into orbiting bands. Earth's are called the Van Allen Belts, after their discoverer. The size of the Van Allen Belts shrink and expand in response to the Solar Wind. The wind is made up of particles, electrons up to 10 Million electron volts (MeV), and protons up to 100 Mev – all ionizing doses. One charged particle can knock thousands of secondary electrons loose from a semiconductor lattice, causing noise, spikes, and current surges. Since memory elements are capacitors, they can be damaged or discharged, essentially changing state.

Galactic Cosmic rays are actually heavy ions, originating outside of our solar system. The actual origin is unknown. They carry massive amounts of energy, up into the billions (10^9) of electron volts.

Radiation Damage to electronics

A complete discussion of the physics of radiation damage to semiconductors is beyond the scope of this book. However, an overview of the subject can be presented. The tolerance of semiconductor devices to radiation must be examined in the light of their damage susceptibility. The problems fall into two broad categories, those caused by cumulative dose, and those transient events caused by very energetic particles, such as those experienced during a period of intense solar flare activity. The unit of absorbed dose of radiation is the *rad*, representing the absorption of 100 ergs of energy per gram of material. A kilo-rad is one thousand rads. At 10k rad, death in humans is almost instantaneous. One hundred kilo-rad is typical in the vicinity of Jupiter's radiation belts. Ten to twenty kilo-rad is typical for spacecraft in low Earth orbit, but the number depends on how much time the spacecraft spends outside the Van Allen belts, which act as a shield by trapping energetic particles.

Absorbed radiation can cause temporary or permanent changes in the semiconductor material. Usually neutrons, being uncharged, do minimal damage, but energetic protons and electrons cause lattice or ionization damage in the material, and resultant parametric changes. For example, the leakage current can increase, or bit states can change. Certain technologies and manufacturing processes are known to produce devices that are less susceptible to damage than others. More expensive substrate materials such as diamond or sapphire help to make the device more tolerant of radiation.

Radiation tolerance of 100 kilo-rad is usually more than adequate for low Earth orbit (LEO) missions that spend most of their life below the shielding of the Van Allen belts. For Polar missions, a higher total dose is expected, from 100k to 1 mega-rad per year. For synchronous, equatorial orbits, that are used by most communication satellites, and some weather satellites, the expected dose is several kilo-rad per year. Finally, for planetary missions to Venus, Mars, Jupiter, Saturn, and beyond, requirements that are even more stringent must be met. For one thing, the missions usually are unique, and the cost of failure is high. For missions towards the Sun, the higher fluence of solar radiation must be taken into account. The larger outer planets, such as Jupiter and Saturn, have their own large, trapped radiation belts around them as well.

Cumulative radiation dose causes a charge trapping in the insulating oxide layers, which manifests as a parametric change in the devices. Total dose effects may be a function of the dose rate, and annealing of the device may occur, especially at elevated temperatures. Annealing refers to the self-healing of radiation-induced defects. This can take minutes to months, and is not applicable for lattice damage. The internal memory or registers of the cpu are the most susceptible area of the chip, and are usually deactivated for operations in a radiation environment. The gross indication of radiation damage is the increased power consumption of the device, and one researcher reported a doubling of the power consumption at failure. In addition, failed devices operate at a lower clock rate, leading to speculation that a key timing parameter was being effected in this case.

Single event upsets (seu's) are the response of the device to direct high energy isotropic flux, such as cosmic rays, or the secondary effects of high energy particles colliding with other matter (such as shielding). Large transient currents may result, causing changes in logic state (bit flips), unforeseen operation, device latch-up, or burnout. The transient currents can be monitored as an indicator of the onset of SEU problems. After SEU, the results on the operation of the processor are unpredictable. Mitigation of problems caused by SEU's involves self-test, memory scrubbing, and forced resets.

The LET (linear energy transfer) is a measure of the incoming particles' delivery of ionizing energy to the device. Latch-up refers to the inadvertent operation of a parasitic SCR (silicon control rectifier), triggered by ionizing radiation. In the area of latch-up, the chip can be made inherently hard due to use of the Epitaxial process for fabrication of the base layer. Even the use of an Epitaxial layer does not guarantee complete freedom from latch-up, however. The next step generally involves a silicon-on- insulator (SOI) or Silicon-on-Sapphire (SOS) approach, where the substrate is totally insulated, and latch-ups are not possible. This is an expensive approach.

In some cases, shielding is effective, because even a few millimeters of aluminum can stop electrons and protons. However, with highly energetic or massive particles (such as alpha particles, helium nuclei), shielding can be counter-productive. When the atoms in the shielding are hit by an energetic particle, a cascade of lower energy, lower mass particles results. These can cause as much or more damage than the original source particle.

The more radiation that the equipment gets, in low doses for a long time, or in high doses for a shorter time, the greater the probability of damage. The Total Ionization Dose (TID) accumulates over time, and actually displaces the semiconductor lattice structure. It causes shifts in the threshold voltage of the device, and noticeable increased current draw. The damage can become permanent. TID is a major concern, as devices become smaller, and the oxide gates become thinner, as technology advances. The higher the voltage, though, the more problematic the effect can be. Analog to digital converters can experience conversion shifts.

These events are caused by high energy particles, usually protons, that disrupt and damage the semiconductor lattice. The effects can be upsets (bit changes) or latch-ups (bit stuck). The damage can "heal" itself, but its often permanent. Most of the problems are caused by energetic solar protons, although galactic cosmic rays are also an issue. Solar activity varies, but is now monitored by sentinel spacecraft, and periods of intensive solar radiation and particle flux can be predicted. Although the Sun is only 8 light minutes away from Earth, the energetic particles travel much slower than light, and we have several days warning. During periods of intense solar activity, Coronal Mass Ejection (CME) events can send massive streams of charged particles outward. These hit the Earth's magnetic field and create a bow wave. The Aurora Borealis or Northern Lights and the Aurora Australias or Southern Lights are manifestations of incoming charged particles hitting the upper reaches of the ionosphere. A series of sun orbiting satellites keeps their eyes on Solar storms, and reports back the Earth. These include Solar orbiter, Solar Maximum,, Solar Radiation and Climate Experiment, Solar and Heliospheric Observatory , Solar Dynamics Observatrory, Hinore (Japan), and others.

The Carrington Event was the most intense geomagmetic event in recorded history, It peaked in early 1859.It seemed to be associated with a bright solar flare. Today, such a storm would wipe out our electrical grids and communications. At the time, they only had the telegraph, which shocked operators. The coronal mass ejection came from the Earth-facing side of the sun.

British amateur astronomers Carrington and Hogdson recorded the first coronal mass ejection event. There were auroral events as south as the Caribbean.

Because of the geomagnetically induced current from the electromagnetic field, telegraph systems all over Europe and North America failed, in some cases giving their operators electric shocks.Telegraph pylons threw sparks.

Cosmic rays, particles and electromagnetic radiation, are omni-directional, and come from extra-solar sources. Some 85%, are protons, with gamma rays and x-rays thrown in the mix. Energy levels range to 10^6 to 10^8 electron volts (eV). These are mostly filtered out by Earth's atmosphere. There is no such mechanism on the Moon, where they reach and interact with the surface material. Solar flux energy's range to several Billion electron volts (Gev).

The effects of radiation on silicon circuits can be mitigated by redundancy, the use of specifically radiation hardened parts, Error Detection and Correction (EDAC) circuitry, and scrubbing techniques. Hardened chips are produced on special insulating substrates. The technology is called silicon on insulator (SOI). Bipolar technology chips can withstand radiation better than CMOS technology chips, at the cost of greatly increased power consumption. Radiation hardened parts are much more expensive than standard parts.

EDAC can be done with hardware or software, but always carries a cost in time and complexity. A longer word than needed for the data item allows for the inclusion of error detecting and correcting codes. The simplest scheme is a parity bit, which can detect single bit (or an odd number of errors, but can't correct anything. EDAC is applied to memory and I/O, particularly the uplink and downlink.

Single Event Upsets (SEU) are instantaneous events, caused by highly energetic particles such as Cosmic Rays. This causes momentary bit flips, but is generally not cumulative. Some events may require a reset to affect recovery. Space Weather id now recolonized as space climate.

Mitigation Techniques

Redundancy can also be applied at the device or box level, with the popular Triple Modular Redundancy (TMR) technique triplicating (make three of everything), and based on the assumption that the probability of a double failure is less than that of a single failure. Watchdog timers are used to reset systems unless they themselves are reset by the software. Of course, the watchdog timer circuitry is also susceptible to failure. Who watches the watcvhdog?

One concept that is easily implemented, and addresses the radiation damage issue, is called Rad Hard software. This is a series of software routines that run in the background on the flight computer, and check for the signs of radiation damage. The biggest indicator is an increase in current draw. The flight cpu will monitor and trend it's current draw, and take critical action such as a reboot if it deems necessary. The Rad Hard software is a variation on self-check routines, but with the ability to take action if needed. We can keep tabs on memory by conducting CRC (cyclic redundancy checks), and one approach to mitigating damage to semiconductor memory is "scrubbing," where we read and write back each memory locations (being careful not to interfere with ongoing operations). This can be done by a background task that is the lowest priority in the system. Watchdog timers are also useful in getting out of a situation such as the Priority Inversion, or just a radiation-induced bit flip. There should be a pre-defined safe mode for the computer as well. Key state data from just before the fault should be telemetered to the control center. Unused portions of memory can be filled with bit patterns that can be monitored for changes. We must be certain that all of the unused interrupt vectors point to a safe area in the code. There is a lot of creative work to be done in this area.

How does Space Weather affect the other planets?

Planets (or moons) with no magnetic field have no defense mechanisms like Earth's Van Allen belts in shielding the planetary surface from Solar emissions. Although the Solar Wind is detectable beyond Pluto, it is far less dense and energetic, and interacts with the interstellar medium.

Mercury, the closest planet to the Sun, gets the full blast. It does have a magnetic field, so a lot of the particles are held at bay. During a Coronal Mass Ejection, Mercury's magnetopause gets pressed down to the surface of the planet. To make it more interesting, Mercury is in tidal lock, so one side always faces the Sun, and the other, deep space. It wiggles a little, so there is a "goldilocks zone," where the temperature might be temperate, and the region has a night-day cycle.

Comets, as they get closer to the Sun, develop a tail, pointing outward from the Sun. The tail consists of material boiling off the comet's surface.

The velocity of the solar wind was observed to be zero at

about 10^7 miles out from Earth, by the Voyager-1 Spacecraft. This is the established boundary of the Sun's influence, beyond which is interstellar space.

Venus

Venus is the hottest planet in out solar system, thanks to the Sun, and it's own heavy cloud layer, made up mostly of sulfuric acid. It has a weak magnetic field. Thr planet is stuck it's own cloud cove.

Mars

Mars has no magnetic field. It has polar ice caps, of frozen water ice, and sections of carbon dioxide. When the poles are in Sunlight, this "snow" or ice sublimates. This action results in clouds, and geysers of CO2 gas. Mars gets less than half the amount of sunlight as does the Earth. When Mars is closest to the Sun, large dust clouds are generated, that can envelope the planet. One of these happened in 2018, affecting the operation of surface rovers. Mars has an observing weather satellite in orbit.

Jupiter

As with Earth, the interaction of Jupiter's magnetosphere with the solar wind generates a bowshock, on the side toward the Sun. On the lee side, the magnetotail extends to the orbit of Saturn. The magnetosphere protects the four largest Jovian Moons. The magnetosphere is a torus around the planet, populated by gas from volcanic activity on the moons. Intense radio waves are generated, and expand out in a cone, which can reach the Earth.

Jupiter has magnificent auroras. The Hubble space telescope keeps track of weather on Jupiter and Saturn

Saturn

Saturn has its own magnetic field, although weaker than Jupiter's by a factor of about 20. This results in the Saturn magnetosphere being smaller as well. It extends out about 20 Saturn radii. The Saturnian moon Titan orbits in the magnetosphere, and gains ionized particles in its atmosphere. Auroras have been seer at Saturn.

Uranus

Uranus' magnetic field is tilted some 60 degrees from the axis of rotation. This results in a great asymmetry, that allows a once-per-day exposure to the solar wind particles. There is a bow shock at 23 Uranian radii sun-wards, and a magnetotail extends millions of kilometers, and forms a corkscrew shape. There are auroras at Uranus as well.

Neptune

Neptune's bow shock is sun-ward, about 35 Neptune radii with the tail extending at least 72 Neptune radii. Not much more is currently known.

Pluto

This minor planet has no magnetic field, so it does not interact with the Solar wind.

How do we observe space weather?

Why, we use satellites in orbit about the Earth, or directly around the Sun. The Earth's atmosphere blocks most to the infrared light emitted from the Sun, but our view from space is much better. We'll discuss a couple of space weather satellites.

SDO

The Solar Dynamics Observatory has been on station since 2010, observing the Sun. It was part of NASA's Living with a Star Program. This sought new knowledge about the Sun-Earth System.
It is operated from GSFC, and is still returning useful data. By studying the sun's emissions, the Science Team is learning more about the Space Weather which influences the Earth.

The spacecraft images the Sun in ten different wavelengths. It has three instruments: the Helioseismic and Magnetic Imager (HMI), the Atmospheric Imaging Assembly (AIA), and the Extreme Ultraviolet Variability Experiment (EVE). HMI gives continuous full disk coverage at a high spatial resolution. AIA takes images of the solar disk in 10 different wavelengths, every 10 seconds. EVE measures the extreme ultraviolet irradiance of the Sun. The instruments take an image of the Sun every second.

Ulysses

Ulysses is a solar polar orbiter, a joint effort between

NASA and ESA in 1990. It collected data until 2009.

The mission was planned to observe the Sun at all latitudes, requiring the spacecraft to leave the plane of the ecliptic. For this, it used a gravity assist at Jupiter. The launch was delayed, as it was originally manifested on Shuttle *Challenger*.

Power was supplied by an RTG instead of solar panels, since the mission did go to Jupiter first, where solar panels are a lot less useful than closer to the sun.

Radio waves generated by plasma were studied with long (72m) antennas. There was also an x-ray detector, two scintillators, and several magnetometers. A scintillator exhibits luminescence – it glows when struck by a particle, making a good particle detector.

The mission was extended several times, as the spacecraft was working well, and valuable data was being obtained. Besides the Sun data, valuable information was obtained at Jupiter, and during the cruise.

Ulysses showed the complexity of the the Sun;s magnetic field; the fact that there was a lot more "space dust" coming into the solar system than thought, and that the Sun's magnetic field was weaker than had been thought.

Parker Solar Probe

The Parker Solar Probe built at JHU-APL for GSFC, has

just launched in a mission to monitor the Sun from a near position of about 8 solar radii. It has a highly eccentric orbit, which allows it to cool off periodically. It will spend 11 days close to the Sun, and 158 days, further out, cooling off. It also has a massive heat shield. The probe will use seven gravity assists over seven years from Venus to establish its orbit at the Sun. This is a maneuver that taps some of the planets energy to slingshot the satellite where you want it.

Parker will approach the Sun to less than 9 solar radii, where it will be in the outer solar corona. It uses small solar panels, some of which retract behind the heat shield. It is the first in being a spacecraft named after a living scientist, a solar astrophysicist. Professor Eugene Parker at University of Chicago. He was 91 years old at launch.

The solar probe concept goes back to a Solar Orbiter project in the 1990's . It was to use a gravity assist from Jupiter, but Parker didn't have to go that far. The Venus flybys will be completed by 2024.

The spacecraft is protected by a solar shield. It will face an intensity of 650 kilowatts per square meter. The shiled, made of reinforced carbon-carbon is 4.5 inches thick. The temperature of the sun-facing side will get to 2,500 degrees F. The shield is covered by a layer of white alumina to reduce heat absorption. It was estimated that the spacecraft would be inoperable withing seconds if the shield were not in place. There are two solar arrays for power. One is retracted behind the shield at 2.5 AU. Another smaller array collects power, and is liquid cooled.

Instruments include direct measurements of magnetic fields, measurements of energetic protons, electrons, and heavy ions, a wide field optical imager, for analysis of solar wind constituents.

Launch and deployment went smoothly, and testing is being conducted. The first science data is expected by the end of 2018.

RHESSI

The Rhessi, the Ramaty High-Energy Solar Spectroscopic Imager, was launched in 2002. It observes in some wavelengths common with the Compton and Chandra missions. It is permanently pointed toward the Sun, and observes solar flares in the hard x-ray through gamma ray range. Rhessi was also the first spacecraft to measure gamma ray flashes from terrestrial thunderstorms. Luckily, its operating life spanned a full eleven year solar cycle. The mission operated until 2018, as this book was being prepared. Its results are being archived, and made available.

Wrap-up

There is weather in Space, coming from the Sun. We have great interest in keeping up with and understanding the Space Weather, since it affects Earth in so many ways. What we learn at Earth will help us to understand the weather at other planets, not just in our own solar system.

Do exo-planets, outside of our solar system, orbiting other stars, have weather? We know they exist, but we have yet to be able to image one. It would be a good bet they do have weather. Wait and see.

Glossary of Terms

ACSWA - American Commercial Space Weather Association.
ASIN - Amazon Standard Inventory Number.
AU – astronomical unit of length – about 150 million km.
Aurora – light display in Earth's atmosphere, caused by Solar wind.
Bow shock – Plasma from the Sun hitting the Earth's magnetosphere.
Carrington Event – (1859) great magnetic storm.
CISM – Center for Integrated Space Weather Modeling.
CME – Coronal Mass Ejections.
CMOS – complementary metal oxide semiconductor.
Corona – Sun's upper atmosphere.
CPU – central processing unit, of a computer.
CSEM – Center for Space Environmental Modeling.
DSCOVR – Deep Space Climate Observatory.
DST – Disturbance storm time Index, strength of the ring current.
EDAC – error detection and correction.
Epitaxial - deposition of a crystalline overlayer on a crystalline substrate.
Erg – unit of work.
ESA – European Space Agency.
EV – electron volt, unit of energy
FAA – (U. S.) Federal Aviation Administration.
GEM – Geospace Environmental Model.
Genesis – NASA solar sample and return mission.
GEO – Geostationary Earth Orbit.
GOES – Geostationary Operational Environmental

Satellite.
Goldilocks Zone – not too hot, not too cold. From a Fairy Tale.
GPO – (U. S.) Government Printing Office.
GSFC – Goddard Space Flight Center.
Heliopause – where the solar wind is stopped by the interstellar medium.
Heliophysics – physics of the Sun.
Heliosphere – plasma from the Sun, detectable beyond Pluto. Area of the Sun's influence.
IGY – International Geophysical Year.
Ionosphere – ionized part of the upper atmosphere, caused by solar radiation.
ISBN – International Standard Book Number.
Isotropic – uniform in all orientations.
ISRO – Indian Space Research organization.
JHU-APL – Johns Hopkins University, Applied Physics Lab.
Joule – SI unit of energy.
Jovian – pertaining to Jupiter
KM – kilometer.
L1 – Lagrange point between two masses. Gravity null.
Lee side – downwind.
LEO – Low Earth Orbit.
Lithosphere – Earth's crust and upper mantle.
LWS – (NASA) Living with a star.
Magnetosheath – volume of high particle energy flux, between the bow shock and the Magnetopause.
Magnetopause – where the planetary magnetic field pressure is in balance with the pressure from the Solar Wind.

Magnetosphere – volume around a body in space with charged particles, held by the body's magnetic field.
Magnetotail – opposite the side of the planet from the solar wind.
MeV – million electron volts.
MMOC - Multi-Mission Operations Center – GSFC.
Moreton wave - large-scale solar coronal shock wave.
NASA – National Aeronautics and Space Administration
NOAA – (U. S.) National Oceanographic and Atmospheric Administration.
NRC – National Research Council.
NSF – (U.S.) National Science foundation.
NSWP – (U. S.) National Space Weather Program.
OGO – Orbiting Geophysical Observatory.
Plasma – ionized gas.
Ring current – current caused by charged particles in the magnetosphere.
SAA – South Atlantic Anomaly.
SDO - Solar Dynamics Explorer.
SEL – Single Event Latchup
SEP – Solar Energetic particles.
SEU – Single Event Upset.
SI – System International; metric
SID – Sudden Atmospheric Disturbance.
SMM - (NASA) Solar Maximum Mission.
SOHO – Solar and Heliospheric Observatory (NASA satellite).
SOI – silicon-on-insulator; technique to mitigate radiation damage.
Solar Wind – charged particles from the Sun's corona.

SPE – solar proton event
SSN – Sunspot number.
STEREO – Solar Terrestrial Relations Observatory. (satellite).
STS – Space Transportation System, Shuttle.
Sun spots – areas on the Sun's surface that are cooler and darker.
TEC – Total Electron Content.
TMR – triple modular redundancy – 3 of everything.
Torus – donut or bagel shape.
USSWP - United States Space Weather Program.

Bibliography

Bothmer, Volker; Daglis, Ioannis A. *Space Weather: Physics and Effects,* 2007, ISBN- 3540239073.

Carlowicz, M. J., and R. E. Lopez, 2002, *Storms from the Sun*, Joseph Henry Press, ISBN-0-309-07642-0.

Daglis, I. A. (Editor), 2001, *Space Storms and Space Weather Hazards*, Springer-Verlag New York, ISBN-1-4020-0031-6.

Eddy, John A. The Sun, the Earth, and Near-Earth Space, NASA, ISBN 978-0-16-08308-8. avail: http://ilwsonline.org/publications/SES_Book_Interactive%20508.pdf

Freeman, John W., 2001, *Storms in Space*, Cambridge University Press, ISBN-0-521-66038-6.

Goodman, John M. *Space Weather & Telecommunications,* 2005, ISBN-0387236708.

Howard, Tim *Space Weather and Coronal Mass Ejections*, 2014, ISBN-1461479746.

Ioannis A. Daglis *Effects of Space Weather on Technology Infrastructure.*, Springer, 2005, ISBN 1-4020-2748-6.

Khazanov, George V. *Space Weather Fundamentals*,

2016, ISBN-1498749070.

Moldwin, Mark *An Introduction to Space Weather,* Cambridge Univ. Press, 2008, ISBN-978-0-521-86149-6.

NASA, *21st Century Complete Guide to Space Weather: Solar Storms, Impacts on Human Activity, Flares and Coronal Mass Ejections, Satellite Sun Observation, Forecasting, Carrington Event*, 2013, ASIN-B00EA1T57E.

NASA, *National Space Weather Action Plan and Strategy: Potential Effects - Power Outages, Infrastructure, Communication, Mitigation Plans, Forecasting, Induced Fields, Solar Radio Bursts,* 2016, ISBN-1973475561.

NASA, *The Sun, the Earth, and Near-Earth Space: A Guide to the Sun-Earth System - Comprehensive Information on the Effects of Space Weather on Human Life, Climate, Spacecraft,* 2013, ISBN-9780160838071. Avail: https://bookstore.gpo.gov/, stock number 033-000-01328-1.

National Research Council, *Severe Space Weather Events: Understanding Societal and Economic Impacts: A Workshop Report,* 2009, ISBN-0309127696.

NOAA, *The National Space Weather Program: Strategic Plan,* 2016, ASIN-B01IAOEPTY.

Poppe, Barbara B.; Jorden, Kristen P. *Sentinels of the Sun: Forecasting Space Weather,* 2006, ISBN-1555663796.

Waheed, Malik Abdul; Khan, Parvaiz Ahmad *Space Weather Phenomena Affecting Geo-Space: The Effects of Sun and Solar Phenomena on The Space Environment of Earth; A Cause-Effect Relationship,* 2017, ISBN-6202074418.

Resources

- https://www.sciencedirect.com/topics/earth-and-planetary-sciences/coronal-mass-ejection
- https://www.ngdc.noaa.gov/stp/GEOMAG/dst.html
- http://www.acswa.us.
- Journal of Space Weather and Space Climate, https://www.swsc-journal.org
- https://spaceweather-submit.agu.org/
- Journal of Space Weather and Space Climate (SWSC),

https://www.researchgate.net/journal/21157251_Journal_of_Space_Weather_and_Space_Climate

- wikipedia, various.

If you enjoyed this book, you might also enjoy one of my other books in the Space series.

Stakem, Patrick H. *16-bit Microprocessors, History and Architecture*, 2013 PRRB Publishing, ISBN-1520210922.

Stakem, Patrick H. *4- and 8-bit Microprocessors, Architecture and History*, 2013, PRRB Publishing, ISBN-152021572X,

Stakem, Patrick H. *Apollo's Computers,* 2014, PRRB Publishing, ISBN-1520215800.

Stakem, Patrick H. *The Architecture and Applications of the ARM Microprocessors,* 2013, PRRB Publishing, ISBN-1520215843.

Stakem, Patrick H. *Earth Rovers: for Exploration and Environmental Monitoring,* 2014, PRRB Publishing, ISBN-152021586X.

Stakem, Patrick H. *Embedded Computer Systems, Volume 1, Introduction and Architecture*, 2013, PRRB Publishing, ISBN-1520215959.

Stakem, Patrick H. *The History of Spacecraft Computers from the V-2 to the Space Station*, 2013, PRRB Publishing, ISBN-1520216181.

Stakem, Patrick H. *Floating Point Computation*, 2013, PRRB Publishing, ISBN-152021619X.

Stakem, Patrick H. *Architecture of Massively Parallel Microprocessor Systems*, 2011, PRRB Publishing, ISBN-1520250061.

Stakem, Patrick H. *Multicore Computer Architecture,* 2014, PRRB Publishing, ISBN-1520241372.

Stakem, Patrick H. *Personal Robots*, 2014, PRRB Publishing, ISBN-1520216254.

Stakem, Patrick H. *RISC Microprocessors, History and Overview,* 2013, PRRB Publishing, ISBN-1520216289.

Stakem, Patrick H. *Robots and Telerobots in Space Application*s, 2011, PRRB Publishing, ISBN-1520210361.

Stakem, Patrick H. *The Saturn Rocket and the Pegasus Missions, 1965,* 2013, PRRB Publishing, ISBN-1520209916.

Stakem, Patrick H. *Microprocessors in Space*, 2011, PRRB Publishing, ISBN-1520216343.

Stakem, Patrick H. Computer *Virtualization and the Cloud*, 2013, PRRB Publishing, ISBN-152021636X.

Stakem, Patrick H. *What's the Worst That Could*

Happen? Bad Assumptions, Ignorance, Failures and Screw-ups in Engineering Projects*, 2014, PRRB Publishing, ISBN-1520207166.

Stakem, Patrick H. *Computer Architecture & Programming of the Intel x86 Family*, 2013, PRRB Publishing, ISBN-1520263724.

Stakem, Patrick H. *The Hardware and Software Architecture of the Transputer*, 2011, PRRB Publishing, ISBN-152020681X.

Stakem, Patrick H. *Mainframes, Computing on Big Iron*, 2015, PRRB Publishing, ISBN- 1520216459.

Stakem, Patrick H. *Spacecraft Control Centers*, 2015, PRRB Publishing, ISBN-1520200617.

Stakem, Patrick H. *Embedded in Space,* 2015, PRRB Publishing, ISBN-1520215916.

Stakem, Patrick H. *A Practitioner's Guide to RISC Microprocessor Architecture*, Wiley-Interscience, 1996, ISBN 0471130184.

Stakem, Patrick H. *Cubesat Engineeering*, PRRB Publishing, 2017, ISBN-1520754019.

Stakem, Patrick H. *Cubesat Operations*, PRRB Publishing, 2017, ISBN-152076717X.

Stakem, Patrick H. *Interplanetary Cubesats*, PRRB Publishing, 2017, ISBN-1520766173.

Stakem, Patrick H. Cubesat Constellations, Clusters, and Swarms, Stakem, PRRB Publishing, 2017, ISBN-1520767544.

Stakem, Patrick H. *Graphics Processing Units, an overview*, 2017, PRRB Publishing, ISBN-1520879695.

Stakem, Patrick H. *Intel Embedded and the Arduino-101, 2017,* PRRB Publishing, ISBN-1520879296.

Stakem, Patrick H. *Orbital Debris, the problem and the mitigation*, 2018, PRRB Publishing, ISBN-*1980466483.*

Stakem, Patrick H. *Manufacturing in Space*, 2018, PRRB Publishing, ISBN-1977076041.

Stakem, Patrick H., *NASA's Ships and Planes*, 2018, PRRB Publishing, ISBN-1977076823.

Stakem, Patrick H. *Space Tourism*, 2018, PRRB Publishing, ISBN-1977073506.

Stakem, Patrick H. *STEM – Data Storage and Communications*, 2018, PRRB Publishing, ISBN-1977073115.

Stakem, Patrick H. *In-Space Robotic Repair and Servicing*, 2018, PRRB Publishing, ISBN-1980478236.

Stakem, Patrick H. *Introducing Weather in the pre-K to 12 Curricula, A Resource Guide for Educators*, 2017, PRRB Publishing, ISBN-1980638241.

Stakem, Patrick H. *Introducing Astronomy in the pre-K to 12 Curricula, A Resource Guide for Educators*, 2017, PRRB Publishing, ISBN-198104065X.

Also available in a Brazilian Portuguese edition, ISBN-1983106127.

Stakem, Patrick H. *Deep Space Gateways, the Moon and Beyond*, 2017, PRRB Publishing, ISBN-1973465701.

Stakem, Patrick H. *Crewed Spacecraft*, 2017, PRRB Publishing, ISBN-1549992406.

Stakem, Patrick H. *Rocketplanes to Spacecraft*, 2017, PRRB Publishing, ISBN-1549992589.

Stakem, Patrick H. *Crewed Space Stations,* 2017, PRRB Publishing, ISBN-1549992228.

Stakem, Patrick H. *,Enviro-bots for STEM: Using Robotics in the pre-K to 12 Curricula, A Resource Guide for Educators,* 2017, PRRB Publishing, ISBN-1549656619.

Stakem, Patrick H. *STEM-Sat, Using Cubesats in the*

pre-K to 12 Curricula, A Resource Guide for Educators, 2017, ISBN-1549656376.

Stakem, Patrick H. *Visiting the NASA Centers, and Locations of Historic Rockets and Spacecraft,* 2107, PRRB Publishing, ISBN-154965120X.

Stakem, Patrick H. *Lunar Orbital Platform-Gateway*, 2018, PRRB Publishing, ISBN-1980498628.

Stakem, Patrick H. Embedded GPU's, 2018, PRRB Publishing, ISBN- 1980476497.

Stakem, Patrick H. *Mobile Cloud Robotics*, 2018, PRRB Publishing, ISBN- 1980488088

Stakem, Patrick H. *Extreme Environment Embedded Systems* 2017, PRRB Publishing, ISBN-1520215967.

Stakem, Patrick H. *What's the Worst, Volume-2*, 2018, ISBN-1981005579.

Stakem, Patrick H., *Spaceports*, 2018, ISBN-1981022287.

Stakem, Patrick H., *Space Launch Vehicles*, 2018, ISBN-1983071773.

Stakem, Patrick H. *Mars*, 2018, ISBN-1983116902.

Stakem, Patrick H. *X-86, 40th Anniversary ed*, 2018,

ISBN-1983189405.

Stakem, Patrick H. *Exploration of the Gas Giants and the Ice Giants, Space Missions to Jupiter, Saturn, Uranus, and Neptune*, 2017, ISBN-1717814506.

Stakem, Patrick H. Rocket Science-101, 2018, ISBN-1977067697.

Stakem, Patrick H. *Lunar Orbiting Platform-Gateway*, 2017, ISBN-1980498628.

Stakem, Patrick H. *Space Weather*, 2018, ISBN-1723904023.

Stakem, Patrick H. *STEM-Engineering Process*, 2017, ISBN-1983196517.

Stakem, Patrick H. *RISC-V in Space*, 2019, ISBN-1796434388 .

Stakem, Patrick H. *Mars Railroad,* 2019, ISBN–1794488243.

Stakem, Patrick H. *Arm in Space*, 2019, ISBN-1099789133.

Stakem, Patrick H. *Exploiting the Moon,* ISBN-978-1091057852.

Stakem, Patrick H. Terraforming, 2018, ISBN-978-

1790308101.

Stakem, Patrick H. *Exploration of the Asteroid Belt, a new approach*, 2018, ISBN-978-1731049841.

Stakem, Patrick H. *Exoplanets,* 2018, ISBN-978-1731385055.

Stakem, Patrick H. *Planetary Defense*, 2018, ISBN-978-1731001207.

Stakem, Patrick H. *Space Telescopes*, 2018, ISBN-978-1728728568.

Stakem, Patrick H. *Riverine Ironclads, Submarines, and Aircraft Carriers of the American Civil War*, 2019, PRRB Publishing.

Stakem, Patrick H. *Submarine Launched Ballistic missiles,* ISBN-978-1088954904.

Stakem, Patrick H., *Space Command, Military in Space,* ISBN-*978-1693005398.*